I Want to Be an Astronaut!

Printed in China

ISBN-13: 978-0-15-352861-3
ISBN-10: 0-15-352861-3

8 9 10   0940   11 10 09

Harcourt
SCHOOL PUBLISHERS

Visit *The Learning Site!* www.harcourtschool.com

Rocket City, U.S.A.

Huntsville, Alabama, is called "Rocket City, U.S.A." because of its contributions to America's space program. Most of the rockets used by the National Aeronautics and Space Administration (NASA) were made there. Huntsville is also the home of the museum at the U.S. Space and Rocket Center (USSRC).

The USSRC is important to Alabama's tourism industry. It is one of the top attractions. At the outdoor park, visitors can walk right up to real space rockets. Inside the museum, they can enjoy hands-on activities that simulate space flight.

These rockets on display at the U.S. Space and Rocket Center are symbols of exploration.

Future astronauts can reach for the stars at the USSRC's Space Camp. Students at Space Camp take part in the same training as real astronauts. They also complete mock space missions.

Redstone Arsenal

The USSRC opened in 1970 on land that once belonged to the Army's Redstone Arsenal. During the 1950s, Wernher von Braun, a German rocket scientist, directed America's first space rocket program at Redstone Arsenal. Von Braun played an important part in the space race with the Soviet Union and in NASA's moon landing program.

Wernher von Braun in 1955 with a model of an early rocket ship design

Space Race

The space race began on October 4, 1957. On that historic day, the Soviet Union launched *Sputnik*—the first satellite to orbit Earth. To orbit is to follow a path around Earth.

The United States government worked hard to compete with the Soviet Union. In 1958, the government set up NASA. Two years later, von Braun moved over to NASA's new Marshall Space Flight Center. The Marshall Space Flight Center is next to the Redstone Arsenal.

Space Firsts

In the years that followed the *Sputnik* launch, a flurry of space "firsts" made history.

Yuri Gagarin Soviet Yuri Gagarin became the first person in space. On April 12, 1961, he blasted off in *Vostok I.* Gagarin spent two hours in flight.

Alan B. Shepard, Jr. On May 5, 1961, Alan B. Shepard, Jr., became the first American in space. During a four-hour delay, he waited in his small capsule. His flight lasted only 15 minutes.

John Glenn On February 20, 1962, John Glenn became the first American to orbit Earth. During his five-hour flight, he completed three full orbits.

1961
Alan Shepard
first American
in space

1969
Neil Armstrong
walks on the
moon

1960　　　　1965　　　　1970　　　　1975

1961
Yuri Gagarin
first person
in space

1962
John Glenn
orbits Earth

Neil Armstrong On July 20, 1969, millions of people around the world watched *Apollo 11*'s moon landing on television. On that day, Neil Armstrong became the first person to step onto the moon. In Armstrong's famous words, "That's one small step for man, one giant leap for mankind."

Sally Ride In 1983, Sally Ride became the first American woman in space. On June 18, she flew into history on the space shuttle *Challenger*.

Mae Jemison In 1992, Mae Jemison became the first African American woman in space. She flew on the space shuttle *Endeavor*.

Ellen Ochoa In 1993, Ellen Ochoa was the first Hispanic woman in space. She flew on the space shuttle *Discovery*.

1983
Sally Ride
first American woman in space

1993
Ellen Ochoa
first Hispanic woman in space

1980   1985   1990   1995

1992
Mae Jemison
first African American woman in space

The Right Stuff: Astronaut Selection

How does someone become an astronaut? Every two years, NASA accepts applications. About a hundred candidates are chosen from thousands who apply. They will train to become either pilot astronauts or mission specialists.

Pilot Astronauts

Pilot astronauts can be pilots or commanders. Pilots fly the space shuttle. Commanders lead the space shuttle's missions.

Mission Specialists

Mission specialists must know how to work everything on the space shuttle. They could go on spacewalks. They might also make repairs or conduct experiments.

Astronaut candidates train on equipment they will use in space.

Astronaut Requirements

Astronauts must meet certain requirements. They must have the right education. Their health must be excellent. They even need to be the right height!

Pilot Astronaut

- ☑ Must be a United States citizen
- ☑ Must have at least a college degree in math, science, or engineering
- ☑ Must have 1,000 hours experience as a jet pilot
- ☑ Must be between 4 feet 9 inches and 6 feet 4 inches tall
- ☑ Must have excellent vision

Mission Specialist

- ☑ Must have at least a college degree in math, physical science, or engineering
- ☑ Must have three years of work experience in science, math, or engineering

How tall are you?

Astronauts can't be too short or too tall. They must fit safely into space vehicles. In the early days of the space program, astronauts could be no taller than 5 feet 11 inches. Just under the limit, John Glenn climbs into the small *Friendship 7* capsule.

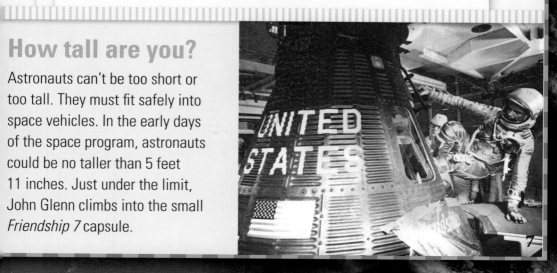

Training for Space

Once chosen, candidates train for two years at the Johnson Space Center near Houston, Texas. They attend classes. They also go through flight and survival training.

In the Classroom...and Beyond

During the first year of training, candidates must take additional classes in science, engineering, and math. They must also learn how to use all the space shuttle's computers and equipment.

Not all training is done in the classroom. Astronauts complete tough physical training, too. On missions, the unexpected often takes place. To prepare, astronauts practice emergency landings. They learn to survive in the woods or in the water.

Astronauts practice first aid while learning to survive in the woods.

The C-9B is a high-tech roller-coaster ride!

Floating in Zero-G

People on Earth take gravity for granted. It is easy to forget that gravity keeps everything from floating away. In space, astronauts work where there is very little or no gravity. Astronauts call this zero-g.

Astronauts train for zero-g in a special airplane, the C-9B. The C-9B flies upward into the sky at high speeds. Then, it suddenly drops. At the top of this high-tech roller-coaster ride, astronauts experience weightlessness. For periods of about 20 to 25 seconds, they float.

Another way to become familiar with weightless conditions is to train underwater. Astronauts work in a tank that is large enough to hold a replica of a space shuttle's cargo bay.

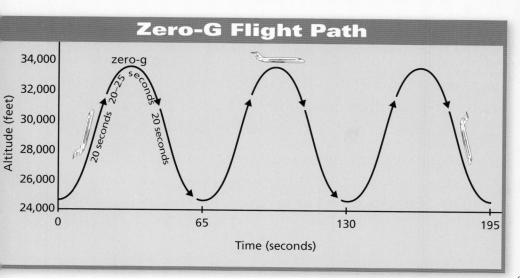

Zero-G Flight Path

Altitude (feet): 34,000 / 32,000 / 30,000 / 28,000 / 26,000 / 24,000

zero-g

20 seconds

20–25 seconds

20 seconds

Time (seconds): 0 / 65 / 130 / 195

Dressed for Success in Zero-G

In everyday life, people sometimes wear special clothes or uniforms. Astronauts wear special clothes, too, depending on what they are doing.

The orange "pumpkin suits," as astronauts call them, are worn during launches and landings. The suit has a special parachute harness. It also holds drinking water and flares.

In case of emergency, these bright suits are easily seen.

While working inside a shuttle or space station, astronauts dress comfortably. They wear T-shirts, sweatsuits, and shorts. These clothes are slightly different from yours, though. They have special fasteners to keep tools from floating away.

Astronauts sometimes must work outside the shuttle or space station. In space, they might make repairs or collect samples. Humans can't live in space. In the sun, it is scorching hot. In the shade, it is bitterly cold. There is also no oxygen, so people can't breathe.

Spacewalkers must wear special suits to protect themselves. The extravehicular mobility unit (EMU) is a suit that has everything an astronaut needs to stay alive in space for many hours.

The EMU's helmet has lights and a TV camera. The visor has a special coating of gold to protect an astronaut from the sun's blinding glare. The cap worn inside the helmet has earphones and a microphone. The backpack holds oxygen for the astronaut to breathe. The manned maneuvering unit (MMU) allows astronauts to walk in space. The MMU is a jet pack that is fueled by nitrogen gas.

EMU Suit

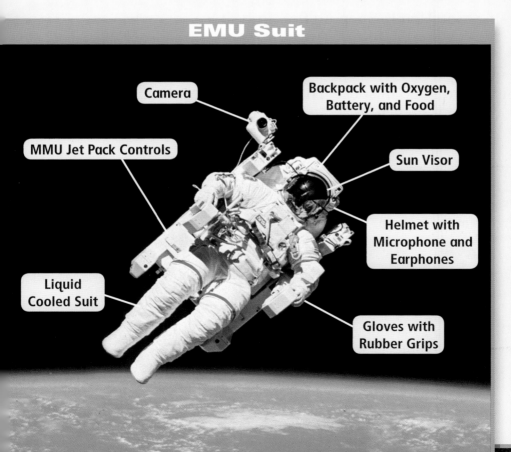

Camera

Backpack with Oxygen, Battery, and Food

MMU Jet Pack Controls

Sun Visor

Helmet with Microphone and Earphones

Liquid Cooled Suit

Gloves with Rubber Grips

Space Camp

Camper tries on a spacesuit.

Would you like to be an astronaut? Since 1982, future astronauts can do more than dream of blasting off one day. They can attend Huntsville's Space Camp!

Rockets Up Close

At Space Camp, the U.S. Space and Rocket Center becomes a large classroom. At the outdoor rocket park, campers can stand right next to a space shuttle. They can see *Freedom 7*, the capsule that Alan B. Shepard flew into space.

Inside the museum, campers can try on space suits. They can taste the astronaut's space food, and find out what it is like to sleep in space. They can squeeze into the cramped Apollo Cockpit Trainer, walk on the surface of the moon, and tour the International Space Station.

Future astronauts arrive at Space Camp.

Students learn teamwork building rockets.

Hey, This IS Rocket Science!

At the beginning of Space Camp, students learn about the science and history of space flight. Then they break into small groups to complete science experiments. As the campers become friends, they learn the value of teamwork.

Some teams build their own model rockets. Other teams design and program robots. At the Outpost-Alpha training area, campers collect soil and rock samples that might be found on the planet Mars.

The campers learn how to use the special tools that astronauts use to work in space. They also visit the Mission Control Room, where they learn about the many people who support the shuttle crew from the ground.

The campers eventually get a chance to show what they have learned. They compete in a game show called the Space Bowl.

Training Like the Pros

Space Campers train like real astronauts. Through simulations, campers experience life in space. They learn to pilot the space shuttle using the controls and switches in the cockpit simulator.

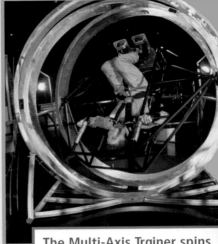

They strap themselves into the Multi-Axis Trainer. The chair in the middle of the machine spins in five directions. It simulates the tumbling that happens during space flight.

The Multi-Axis Trainer spins campers in all directions.

Next, campers climb into the One-Sixth Gravity Chair. It is called the One-Sixth Gravity Chair because the pull of the moon's gravity is one-sixth that of Earth's gravity.

Campers learn to walk in space in a manned maneuvering unit (MMU). The MMU jet pack lets them walk outside without being attached to the space shuttle.

Golfing on the Moon

In 1971, Alan Shepard became the first person to golf on the moon. He took a short one-handed swing. Because the moon's gravity is one-sixth of Earth's, the ball traveled between 200 and 400 yards. The best golf professionals drive the ball a little more than 300 yards.

Suited Up, Strapped In, and Ready to Blast Off

On their second-to-last day of camp, campers complete a test mission. This flight is only one hour long. Counselors tell the campers exactly what to do and say.

Future astronauts complete their mission.

The Final Day

Space Camp's final day is the ultimate test. Campers blast off on a six-hour mission. This time, they all depend on what they have learned at Space Camp. Like real astronauts, they will face the unknown.

During the mission, two emergencies are thrown their way. Will one be rough weather? Mechanical troubles? Perhaps food supplies are running low. Now, campers must use all they have learned, think through problems, and survive in space.

After completing their mission, the campers graduate outside next to the space shuttle. Maybe one day they will become real astronauts.

Think and Respond

1. Why is Huntsville, Alabama, called "Rocket City, U.S.A."?

2. Who was the first American to land on the moon?

3. What are some activities at Space Camp?

4. How does the astronauts' training differ from Space Camp training?

5. What do you think is the most valuable experience at Space Camp? Why?

 # Activity

Draw your own space vehicle. Make a list of the crew and equipment you would need on the first mission in your vehicle.